The Bell Bowl Prairie in Rockford, Illinois. Destruction of an Ecosystem and a Social Identity Motivated by Economics

Leo Kempe

Bibliographic information published by the German National Library:

The German National Library lists this publication in the National Bibliography; detailed bibliographic data are available on the Internet at http://dnb.dnb.de.

ISBN: 9783346999863

This book is also available as an ebook.

University of Illinois at Urbana-Champaign   Leo Kempe

Spring Term 2023

The Bell Bowl Prairie in Rockford, Illinois – Destruction of an Ecosystem and a Social Identity Motivated by Economics

Date of Submission: May 7, 2023

Table of Contents

Abstract

Back in the day, the tallgrass and other kinds of prairies were an integral part of the physical geographic features of the United States, more exactly the Midwest and the South. A vast area was covered with this landscape type, providing a habitat for plants and animals as well as an important mediator for agricultural relations between humans and nature. However, in the last few decades, economic growth models along with increasing population size and density have begun to threaten this vulnerable ecosystem. The most recent evidence of this loss of landscape is the expansion plans of the Chicago Rockford International Airport, pushed forward by the Greater Rockford Airport Authority (GRAA) and threatening the adjacent Bell Bowl Prairie. After a long legal stand-off and official writings, the Federal Aviation Administration (FAA) has approved the destructive endeavor. Looking at this process as an outsider reveals important information about the entire project, and it is argued that this contributed to a loss of social identity for the residents and geographic cohesion for this area.

1. Introduction: Prairie in the State of Illinois

Like other Midwest and Southern states, the 'Land of Lincoln' (Tinus 2023) was largely dominated by several different types of prairies already thousands of years ago, with a wide geographical variation. Indeed, so much so that around 1842, the moniker of the 'Prairie State' was coined (Joliet School District 86 2017; Tinus 2023; Only In Your State n.d.; NLI 2021a: 2). 'Prairie' is derived from a French term meaning a meadow grazed by cattle (Illinois Natural History Survey n.d.). Since the settlers' time, the landscape forms an important part of Illinois' geographical/environmental history and the social identity of its inhabitants: Tinus (2023) writes that Illinoisans have marked their calendars for one week of appreciation of this precious ecosystem and its physiological features in the month of September. Several items of media coverage and symposia have been dedicated to the landscape as well (Illinois Natural History Survey n.d.).

To denote the significance of the prairie and the substantial soybean and corn fields, the 'Garden of the West' is another, though less commonly used nickname for the state (Tinus 2023). Like the more common one, it should capture the seemingly unbreakable power of the prairie, both in terms of its mere presence and its ability to gradually evolve into an identity marker. Hence, rhetoric itself and then circulation inside popular discourse was an important contributor to previous conservation and preservation efforts. Moreover, the region in which most of the old prairie is located is called the "prairie peninsula" (Illinois Natural History Survey n.d.). This covers the entire state of Iowa, substantial parts of the Dakotas, Nebraska, Wisconsin, Missouri, Kansas, and Minnesota, as well as smaller patches of Indiana, Kentucky, Oklahoma, and Arkansas (ibid.).

Therefore, understandably, attempts at the unhindered destruction of the pristine nature contained within the prairie cause visible discomfort and opposition among most Illinoisans. Interestingly, this sentiment covers a wide variety of groups in the Illinoisan population, from conservation and environmental activist institutions to governmental agencies to student organizations such as Red Bison, The Wildlife Society at the University of Illinois at Urbana-Champaign (TWS UIUC), and the Illini Wildlife and Conservation Club (IWCC), also at this university.

Opposing views are especially, but not exclusively, detectable in the counties in which parcels of prairie have remained to this day. One of these counties is Winnebago County in the northwestern portion of the state, with Rockford as a dominant economic center (Library of Congress n.d.). A small but nevertheless significant portion of the ancient prairies in Illinois, Bell Bowl, is located here, and it was long set out to destruction by an administrative body for airplane operations at the local airport. Evidently, this was material for a contestation among several actors with diverging opinions on the matter.

This paper analyzes the legal and social process of negotiation and activism in the Bell Bowl case, which involves a variety of stakeholders on both sides of the aisle. It argues, first, that Bell Bowl Prairie is indispensable for a healthy ecosystem network in Illinois, and second that the very institutions entitled to protect nature undermined Illinoisan preservationists' efforts to retain both the prairie and an important pillar of their identity as residents of this state. After outlining the natural and historical characteristics of the area, the paper proceeds to elaborate on the economic motives for its destruction and the goals of said preservationists. In the third step, legal proceedings and community opposition are explained, using a variety of governmental and non-profit public documentation. To prove the paper's continuous relevance, the current state is also briefly reviewed. Finally, the paper provides a conclusion of findings and the argument as well as an outlook on the dualistic 'economy versus nature' notion. To encourage further research and detailed thoughts, the references are provided in a bibliography as well.

2. The Case of the Bell Bowl Prairie

The Bell Bowl Prairie is an integral part of the landscape in the Greater Rockford area northwest of the City of Chicago. For thousands of years, it has provided cover and nutrition to animals as well as a habitat for typical plants. In fact, it represents one of the state's most biodiverse areas in which a substantial number of endemic species find their environmental niche. The first subchapter is going to explain the characteristics of the Bell Bowl. This helps to set clear geographical and biological boundaries of the area and fosters an understanding of the local relevance of the case illustrated in coming subchapters. Elucidating some spatial facts for contextual purposes is indispensable here as readers might not be familiar with the landscape.

2.1. Characteristics

Unlike the probable first associations with the term 'prairie', Bell Bowl is not a wide, open landscape that is solely covered by high, thin grass gently swinging in the wind. Instead, it is "a little island" (White 2021a: 2) originally covering 25 acres, classified according to composition and status as a so-called "dry gravel remnant prairie" (Save Bell Bowl Prairie n.d.), consisting of "ground-up limestone or dolomite bedrock" (White 2021b: 3). This is a key reason why the space was not converted into agricultural land (White 2021a: 2). Gravel is "loose rounded fragments of rock" (Merriam-Webster n.d.), alluding to specific geological, thermal, and hydrological events that have formed this landscape over thousands of years.

Indeed, like other types of prairies in Illinois and other states, gravel prairie emerged because of the varying scope and scale of "soil moisture, soil composition, geological substrate, glacial history and topography" (Illinois Department of Natural Resources n.d.). Geological substrate means "a surface or volume of sediment or rock where physical, chemical, and biological processes occur, such as the movement and deposition of sediment, the formation of bedforms, and the attachment, burrowing, feeding, reproduction, and sheltering of organisms" (Valentine 2019: 4). In other words, it is a significant component of a habitat. Kerry Leigh, Executive Director of the Natural Land Institute (NLI), a "member-supported, not-for-profit organization" (NLI 2021a: 4), says that Bell Bowl has been around since the retreat of the last Illinoisan glaciers about 10,000 years ago (Kranking 2022).

Because this habitat is quite dry in its majority, only species that have adapted to these conditions can live here. Gravel prairies have high amounts of calcium stored in the rocks themselves and the soil underneath them, permitting plants that do not need a wide variety of nutrients to grow there. These plants, in turn, can provide the necessary nutrition base for birds and insects, being responsible for the further distribution of seeds. Not only the intrinsic conservation of their own species through these mechanisms is possible, but plants can prevent soil erosion and things such as acidification or salinization through their root networks (White 2021a: 5).

In Bell Bowl at Rockford, specifically, the landscape is a mixture of grassland and ground covered in gravel. Suarez (n.d.) says that this place not only provides a thriving space to live for numerous rare species but also contributes to "creating wellbeing for the thousands of birders and prairie enthusiasts across the state" (Suarez n.d.). Hence, it is a geographical space that has significance for the identity of a substantial fraction of the population. These people come out to enjoy the natural space and might collect their own individual data on this vulnerable ecosystem which is on its way to declining further. Not only might this knowledge acquisition trigger more conversation, information exchange, and environmental awareness, but it might also enhance feelings of belonging among the people of the Prairie State.

As to the species at Bell Bowl, Suarez alerts us to the "fragmented landscape" that is also discussed further throughout this paper. The United States Fish and Wildlife Service (USFWS) had to remove several birds and other animals from the endangered species list to officially declare them extinct, making clear how serious the reality of widespread biodiversity and habitat loss has become over the last few years. Many of the birds that live at Bell Bowl, Suarez explains, build their nests on the ground, and when larger populations occur, they need more nesting space. Hence, a prairie that covers a wide area is ideal, even more so when many dense plants grow that shield the birds from predators.

Further, higher plant diversity is found in prairies like Bell Bowl, which is why rather rare insects, small mammals, or fungi can flourish in such an environment. Birds, who are often migratory, can benefit from this variety to stock their reserves of minerals and other nutrients. Many of these bird species' numbers have been declining rapidly over the last years and decades, a reality that is prospected to be even more pronounced soon. As mentioned, many of the birds are migratory, meaning they only live in places like Bell Bowl for their breeding season. These birds include "the state endangered Loggerhead Shrike and state threatened Black-billed Cuckoo, as well as Bobolinks, Bell's Vireos, and 90 other species." The number has been retrieved from the eBird database by Kranking (2022) for her article. Hence, Illinois prairies are indispensable for the longevity of these species – if not protected, these species will have less fortune to live.

2.2. The Old Controversy: Economy versus Environment

Over the last few decades, our society has pursued a resource-intensive and consumerist agenda that continuously threatens all kinds of habitat types, from forests to lakes and rivers to mountain ranges and the open ocean. The desire for more goods as well as population growth has influenced our economy to the extent that we need more and more performative machines to produce more and bigger amounts of food and convenience goods. Further, we crave more natural materials to satisfy this greed, such as minerals, coal, water, fruit, or meat. Entire ecosystems, such as the prairie, have suffered because people fundamentally changed their lifestyle choices – particularly when they move to the cities.

Urbanization and large-scale compact development have long been in conflict with concepts of 'wilderness' and 'countryside' because many saw urbanites as reckless in relationship to nature, while they themselves might have felt an infringement on their personal choices by an exaggerated 'naturalist' agenda. Because of this notion crystallizing over the years and decades of our history, the reconciliation of those two interests became increasingly difficult. Urban dwellers often view natural spaces as a distortion of a thriving economy, a perfectly networked structure, and spheres in which windows of opportunity open for success. Nature, by contrast, customarily represents the unknown, the irrational, the wide, and the empty. Many urbanites do not see the richness of the natural environment that people like birders, rangers, or hunters would see.

This is where the notion of sustainability comes into play: How can cities and other human living spaces be developed or managed in a way that does not completely disregard the non-human world? In the political world, several organizations have tried to find solutions that attempt to reconcile human desires and the cravings of our natural surroundings. The United Nations Department of Social and Economic Affairs (UNDESA), for example, presents the 17 Sustainable Development Goals (SDGs) on its website. Together with their indicators, they are used to assess the necessities of people and nature and thus as guiding principles for more substantial policy decisions. 'Human' goals are SDG 1 (No Poverty), 2 (Zero Hunger), 3 (Good Health and Wellbeing) or 6 (Clean Water and Sanitation). Most SDGs encompass strictly human factors (food, water, equality), whereas Goal 8 (Decent Work and Economic Growth) or Goal 9 (Industry, Innovation and Infrastructure) are more economic. Goals 13 (Climate Action), 14 (Life Below Water), and 15 (Life on Land) look decisively at nature. However, even though it might not seem so at first glance, humans are generally a part of all goals.

Nonetheless, it is two worlds colliding, worlds that despite Carolyn Finney's description of the entire urban fabric as nature are fundamentally different. The dualism of the subchapter's title is equally visible with the Bell Bowl Prairie. It is located adjacent to one of Illinois' most significant airports, the Chicago Rockford International Airport, being a hub for international air cargo operations as well as a few commercial airlines with several national destinations (RFD n.d.). The Greater Rockford Airport Authority (GRAA), the body administering the airport, has announced thorough plans for the expansion of this property to allow the construction of additional storage space as well as other parts of the whole technical facility (NLI 2021a: 6). Evidently, this would increase the airport's regional, national, and international position and prominence but would go to the detriment of the sensitive ecosystem nearby.

Equal to other projects, the fractions of society in the Bell Bowl case have separated into conservationists and preservationists on the pro-environment side of the aisle, and pro-development, pro-profit agencies and organizations on the other. However, inside this 'environmentalism', the former is distinct from the latter: Conservation means that while the human-nature relationship should be sustainable, natural resources will still be extracted. If policy measures for driving down resource use are beneficial for humans, conservationists will support those. They do not want to give up economic growth for seemingly too much protection of nature (Smith 2018; Our Endangered World 2021). They support the notion that resource extraction is targeted at humans and their unfettered development over years, decades, and centuries.

Preservationists, on the other hand, typically take a more 'radical' approach to the protection of valuable resources and landscapes: They aim to protect a certain geographical space where there are no humans, human-made construction (e.g., roads), or – if there are humans – if they are native. Hence, the diligent protection of some natural areas is often justified on religious and spiritual grounds as well (Smith 2018). The article by Our Endangered World, unfortunately, not only mixes up the concepts in its explanation but describes the same concept in diverging terms. This creates confusion among readers.

The next subchapter describes the processes of legal opposition against the decisions taken by both the GRAA and federal agencies such as the Federal Aviation Administration (FAA) or the United States Department of the Interior.

2.3. Legal Proceedings and Community Opposition

Although one might think that the dispute about this vulnerable ecosystem in the Rockford area is new, it has been going on for many years and decades even before the official expansion announcement was made. In fact, it was a man by the name of George B. Fell – the founder of NLI – who negotiated with the GRAA about the protective status of the prairie back in 1968 (White 2021a: 2; White 2021b: 6). Obviously, this paper cannot go back into the more distant history of the socio-political and legal ramifications that were already recorded. However, it nevertheless proves valuable in a different context because it is important to see that for quite some time, the area has been on the radar of the news and environmental policymaking.

One of the big parties in the pro-environmental camp was the Natural Land Institute (NLI), an organization that fought tirelessly for Bell Bowl since the official announcement made by the GRAA. This organization will be a topic in the next few paragraphs. However, also individual people like John White, of Urbana, were among the earliest Bell Bowl advocates. White is the designer and director of a six-category natural area classification and overall guidance framework known as the Illinois Natural Areas Inventory or INAI. According to the Illinois Department of Natural Resources Division of Natural Heritage, INAI started in 1978 and contains "ecological information on natural areas evaluated to have statewide conservation significance" (IDNR Division of Natural Heritage n.d.; White 2021b: 8). Interestingly, the website then speaks of the INAI's ability to encourage the effective *preservation* of wildlife and plants, indicating that the difference explained before is perhaps not made in this explanation. Bell Bowl Prairie analyzed here is numbered as 0916 within Winnebago County and falls into Category I – High quality natural community and natural community restorations (Illinois Natural Heritage Database 2023; NLI 2021a: 20; White 2021b: 7).

In his statement of September 23, 2021, White not only described his environmentalist work in the United States and other countries but particularly stressed the immense significance of Bell Bowl Prairie as an ecosystem (White 2021a). He alerted the representatives of the GRAA to the "marvelously intricate ecology" (1) and the "rare treasure" (ibid.) that Bell Bowl represents in the natural composition of the State of Illinois. When he says that the "prairie still exists because of the airport" (ibid.), he connects Bell Bowl Prairie to the ramifications for the residents' social identity in case of the forceful destruction of this natural space. The GRAA had officially committed to an efficient management practice of the prairie space back in 1977 through a Master Plan and Resolution (White 2021a: 3; NLI 2021a: 18 f.).

However, because of the plans, White demanded redirection of developmental efforts as well as corresponding arrangements to further protect the prairie's natural integrity. On the remaining pages, White (2021a) makes a strong and botanically well-informed case about why it would be the absolutely wrong decision to move Bell Bowl around. He talks about the uprooting of plants, the disturbance in general, as well as the reality that through adaptation to disturbed conditions, native prairie plants would be eliminated from their habitat through ecological competition with other plant types.

As mentioned, the NLI was a prominent actor in the debate about what would happen to Bell Bowl because its employees have managed it for decades (see NLI 2021a: 19). On October 26, 2021, the long process began with the NLI filing a Temporary Restraining Order (TRO) to bring the GRAA to think differently about its expansion of the Midfield Cargo Development proposed in 2019 (see NLI 2021a: 20). Besides the GRAA, other agencies such as the United States Department of Transportation (US-DOT) and the U.S. Fish and Wildlife Service (USFWS) were among the Defendants as well. The NLI sought both "declaratory and injunctive relief" (NLI 2021a: 2), that is, the possibility to prevent future damage that could be avoided by alternative conduct. The former requires stating the existence of a controversy, "a connection between the challenged conduct and injury, and redressability that the court could order" (Legal Information Institute n.d. a) but does not result in actions that can be enforced against defendants. Therefore, the seemingly stronger remedy of injunction was sought as well, something that requires an involved party to pursue certain actions (ibid. b)

In particular, the NLI stresses that several species living in Bell Bowl are legally classified as endangered by the Endangered Species Act (ESA), such as the plant species Prairie Dandelion and Large-Flowered Beard Tongue (LFBT) and the Rusty Patched Bumble Bee (RPBB) (NLI 2021a: 2, 3, 20, 28). The Plaintiff reproaches the Defendants for having actively violated statutes on both the state and federal level, a reality that they do not tolerate under any circumstances. The same goes for the 1977 agreement in which the GRAA ensured to provide some developmental alternatives and reliable management practices that would not impact the area covered by Bell Bowl.

Furthermore, environmental assessments issued in 2019 framed the matter as a Finding of No Significant Impact, or FONSI, a document signed to approve the expansion. However, this was *before* visual evidence of the RPBB was collected and the site was allegedly removed from the INAI list according to that text (NLI 2021a: 21 f., 29 ff.). Because of this, the Illinois Nature Preserves Commission (INPC), Illinois Department of Transportation (IDOT), and Illinois Department of Natural Resources (IDNR) had to enter new consultations as well as email and letter communication (see NLI 2021a: 22f.). On pages 22 and 23 of this first document, the authorities use quite a plain language regarding what would happen to Bell Bowl: The place would be "'slated for destruction, potentially in the coming days'" (NLI 2021a: 23) or that "'much of the Bell Bowl Prairie INAI site will be destroyed'" (ibid.).

All relevant agencies also state in this communication that all alternatives shall be sought and considered, thus suggesting an unclear position as to what should happen (ibid.): preservation or destruction. However, further email communication well into September 2021 detailed within the filed TRO reveals major communication issues within all involved federal and state agencies, ultimately leading to the impression of undermining protective efforts and successful negotiations. It is also explained that the GRAA for itself was reluctant from the start to consider developmental alternatives for its planned project impacting the ecosystem at Bell Bowl. Apparently, a "'landowner has every right to do whatever they want to an INAI'" (NLI 2021: 26) and that it is unlikely that the airport will conduct a reconsideration of the issue. On pages 34 ff., several detailed violations reproached by the NLI are listed by the respective law and statute. NLI states severe communication and approval request failures by the Defendants, leading to legal and social chaos, as well as the loss of natural integrity and social significance.

Although the TRO was withdrawn in a later motion, the NLI continued to battle GRAA et al. in many subsequent cases challenging the authority, integrity, reliability, and accountability of the GRAA as well as the other state and federal agencies. To prolong the time of negotiation, the NLI sent a letter for a sixty-day notice of the violation of ESA Section 7 to the Assistant United States Attorney for the Northern District of Illinois (NDIL), Monica V. Mallory (Russell 2021: 8, 10, 12 ff.). The letter, much like the TRO, details the failures of national-level agencies, particularly the Federal Aviation Administration (FAA). One of the major issues discussed is that these agencies "failed to ensure their funding, authorization, construction, expansion and operation of RFD does not jeopardize the continued existence of the Bee" (ibid.). Russell says that not only did the agencies ignore the actual existence of the RPBB, but they knowingly did so, while additionally evading the requirements of several supplemental documents elaborating on the natural condition as well as the direct and indirect effects of the expansion activity. Evidently, this creates trust issues and affects Bell Bowl as natural and identity marker.

During late 2021 and early 2022, the legal stand-off had escalated into GRAA claims that the NLI does not have right to be on the airport property and cannot claim the injunctive relief initially filed for. In essence, the GRAA along with the federal agencies said that the NLI does not have the right to challenge the activities leading to the expansion of the airport premises. In other words, the GRAA reproached the Plaintiff for unlawfully stating claims about the expansion activities even though they allegedly do not have any jurisdictional power in matter.

2.4. Current Status

In late 2022, the USFWS concurred with the FAA's determination of no significant effect on the RPBB or other animal and plant species inside the prospected construction area, according to a letter by USFWS Illinois-Iowa Field Office Supervisor Kraig McPeek (2022). While mentioning the existence and sighting of the RPBB on-site, several controlling activities are proposed on pages 2 ff., such as no usage of insecticides. The claims is that the Biological Assessment (BA) presented to the agency "considers the appropriate statutory and regulatory factors and is supported by the best available scientific data" (McPeek 2022: 4). However, generally, this suggests that the USFWS is not willing to consider preservationists' opinions and strong concerns.

Furthermore, the Illinois Department of Natural Resources has closed consultations with a letter written by Brayden Hayes, the Manager of the Impact Assessment Section within the Division of Real Estate Services and Consultation. Although the letter states that "closed consultation does not imply the Department's authorization or endorsement of the proposed action", it suggests the IDNR nevertheless supports the decision of the FAA and GRAA. The first paragraph says that several species were evaluated but no significant impact found regarding the proposed actions such as shrub and brush clearing.

Following this development, the NLI has filed petitions for review respectively emergency motions for review of the Written Reevaluation by the FAA, both of which are targeted to halt construction activity. The Emergency Motion for Stay Pending Review (2023) states that after "arbitrary and capricious federal reviews, assessments and findings, the FAA has now authorized the GRAA – by virtue of its March 3, 2023 Written Reevaluation (on which NLI's Petition for Review, filed March 4, 2023, is based)—to bulldoze Bell Bowl Prairie as early as Thursday, March 9, 2023" (NLI v. Dickson 2023: 1; Illinois Environmental Council 2023).

3. Conclusion and Outlook

The Illinoisan prairie is one of the treasures of the ecosystem diversity of planet Earth. It has been there for thousands of years, unchanged by human activity. However, the Anthropocene has changed this reality fundamentally.

After a long, hurtful legal battle, the destruction of the Bell Bowl Prairie south of the City of Rockford is a fact. The Natural Land Institute, along with several other environmental advocates, has fought to preserve the habitat of some of the rarest species occurring in the State of Illinois, ultimately succumbing to the pressures of the Greater Rockford Airport Authority expansion plans. This is a severe shock not only to the environmental and natural integrity of the area, but also a loss of a social identity marker for the residents. Many people have cherished the prairie over the years as a place to experience nature and to study all the marvelous species that occur here.

The challenge to understand the significance of the prairie and the Bell Bowl case in particular demands substantial further social and legal research. The paper could not go into the small-scale details regarding the violation of acts or the communication failures that have occurred, but it is necessary to provide an analysis as detailed as possible. Separate papers could incorporate further insights into communication patterns in the relationship of state and federal agencies as well as the legal nature of specific laws of relevance in the case.

Overall, the fate of Bell Bowl has been decided by an economic instead of an environmental force, and it is the indispensable task of researchers and activists to stand their ground in openly debating and writing about these issues. As I am a strong advocate of environmental protection efforts, this would relate to further career opportunities in landscape planning, physical geography, and human geography.

Reference List

Chicago Rockford International Airport (n.d.). "Who we are." RFD, online at:
https://flyrfd.com/who-we-are/.

Chicago Rockford International Airport (n.d.). "Environmental." RFD, online at:
https://flyrfd.com/environmental/.

DeCoster, K. (2023). "Construction resumes at Bell Bowl Prairie near Rockford airport."
Rockford Register Star, online at:
https://www.rrstar.com/story/news/local/2023/03/09/construction-resumes-at-bell-
bowl-prairie-near-rockford-airport/69989318007/.

Federal Aviation Administration (2023). "Written Reevaluation for the 2019 Environmental
Assessment for the Northwest Cargo and Midfield Development Area and the 2021
Categorical Exclusion for the South Cargo Development Area - Air Cargo Facility &
Associated Improvements." FAA, available at: https://drive.google.com/file/d/1-
94cfyARVxN9K1wA89j1KK9VTlfATd9r/view.

Finney, C. (2013). "Ode To New York: A Performance Piece." Humans Nature, online at:
https://humansandnature.org/urban-land-ethic-carolyn-finney/.

Grand Prairie Friends (n.d.). "What We Do" GPF, online at:
https://www.grandprairiefriends.org/what-we-do.

Haas, K. (2021). "Rockford Airport Says Roadway Can't Be Redesigned to Avoid Bell Bowl
Prairie." Rock River Current, online at:
https://www.rockrivercurrent.com/2021/11/10/rockford-airport-says-roadway-cant-be-
redesigned-to-avoid-bell-bowl-prairie/.

Illinois Department of Natural Resources (n.d.). "Illinois Natural Areas Inventory."
Illinois.gov, Illinois Department of Natural Resources (IDNR), Division of Natural
Heritage. Online at: https://naturalheritage.illinois.gov/naturalareasdivisions/illinois-
natural-areas-inventory.html.

Illinois Department of Natural Resources (2023). "Illinois Natural Areas Inventory (INAI)
sites by County." IDNR, Illinois Natural Heritage Database, available at:
https://naturalheritage.illinois.gov/content/dam/soi/en/web/naturalheritage/dataresearc
h/documents/INAICountyList%20apr2023.pdf.

Illinois Department of Natural Resources (2022). "Transmittal of Presence/Absence Surveys
for Midfield Development Project at Chicago Rockford International Airport,

Rockford, Illinois." IDNR, online at:

    https://drive.google.com/file/d/1Pf0_XOTBfCqtpaG--WAjmeSpHDFK8AEl/view.

Illinois Environmental Council (2023). "Priceless Bell Bowl Prairie Demolished In Rockford,

    IL." IEC Press Release, online at: https://ilenviro.org/priceless-bell-bowl-prairie-

    demolished-in-rockford-il/.

Illinois Natural History Survey (n.d.). "Tallgrass Prairie." INHS, online at:

    https://publish.illinois.edu/tallgrass-prairie/.

Joliet School District 86 (2017). "Illinois: The Prairie State." Available at:

    https://www.joliet86.org/assets/1/6/Day_1_Reading_Illinois_article_and_quiz.pdf.

Kranking, C. (2022). "Bell Bowl Prairie Activists Stay Steadfast as a Rare Habitat Remains in

    Limbo." Audubon Great Lakes, National Audubon Society, available at:

    https://www.audubon.org/news/bell-bowl-prairie-activists-stay-steadfast-rare-habitat-

    remains-limbo.

Legal Information Institute (n.d.) a. "Declaratory relief." LII, Cornell Law School. Online at:

    https://www.law.cornell.edu/wex/declaratory_relief.

Legal Information Institute (n.d.) b. "Injunctive Relief." LII, Cornell Law School. Online at:

    https://www.law.cornell.edu/wex/injunctive_relief.

McPeek (2022). "Concurrence of Biological Assessment." USFWS Illinois-Iowa Field Office,

    online at:

    https://drive.google.com/file/d/1LNqF9hjNnRcW2ohqstbCoOTdBYCl1vUn/view.

Merriam-Webster Dictionary (n.d.). "Gravel." Available at: https://www.merriam-

    webster.com/dictionary/gravel.

Natural Land Institute (n.d.). "Bell Bowl Prairie." NLI, available at:

    https://www.naturalland.org/bell-bowl-prairie/.

Natural Land Institute (2023). "Natural Land Institute Statement about Temporarily Halting

    Development and Construction on Bell Bowl Prairie." Media Release, online at:

    https://www.naturalland.org/wp-content/uploads/2021/10/Media-Release_NLI-

    Statement_Bell-Bowl-Prairie-10.28.21.pdf.

Natural Land Institute (2021) b. "Defendants' Motion to Dismiss for Lack of Standing."

    United States District Court, Northern District of Illinois, Western Division. Case No.

    3:21-cv-50410. Available at: https://www.naturalland.org/wp-

    content/uploads/2021/12/Filed-Defendants-Motion-to-Dismiss-for-Lack-of-

    Standing_12.17.21.pdf.

Natural Land Institute (2021) a. "Plaintiff's Complaint for Declaratory and Injunctive Relief."
United States District Court, Northern District of Illinois, Western Division. Case No.
3:21-cv-50410, Temporary Restraining Order (TRO), available at:
https://www.naturalland.org/wp-content/uploads/2021/10/NLI-v.-GRAA-et-al.-Case-
No.-21-cv-50410.pdf.

Library of Congress/LoC (n.d.). "History of Winnebago County, IL." Available at
https://tile.loc.gov/storage-
services/public/gdcmassbookdig/historyofrockfor00chu/historyofrockfor00chu.pdf.

Only In Your State (n.d.). "8 Ways Illinois Got The Nickname 'The Prairie State.'" Available
at: https://www.onlyinyourstate.com/illinois/prairie-state-il/.

Our Endangered World (2021). "Conservation vs. Preservation: The Differences. Available
at: https://www.ourendangeredworld.com/eco/conservation-vs-preservation/.

Russell, J.M. (2021). "Sixty-Day Notice of Violations of the Endangered Species Act relating
to the Chicago Rockford International Airport's Expansion at Bell Bowl Prairie,
Winnebago County, Illinois." Von Briesen & Roper, available at:
https://www.naturalland.org/wp-content/uploads/2021/12/NOI-Letter_J.-Russell-Ltr.-
to-M.-Mallory-2021.11.19.pdf.

Save Bell Bowl Prairie (n.d.). "Resources." Available at:
https://www.savebellbowlprairie.org/resources.

Smith, A. (2018). "What's the Difference Between 'Conservation' and 'Preservation'?
Piedmont Environmental Alliance, online at: https://www.peanc.org/whats-difference-
between-conservation-and-preservation.

Suarez, D. (n.d.). "Bell Bowl Prairie Provides Habitat for Rare and Threatened Species of
Wildlife and Nature for Birders Across Illinois." National Audubon Society, available
at: https://gl.audubon.org/news/bell-bowl-prairie-provides-habitat-rare-and-threatened-
species-wildlife-and-nature-birders.

Tinus, D. (2023). "Why Is Illinois Called the Prairie State?" UnitedStatesNow. Available at:
https://www.unitedstatesnow.org/why-is-illinois-called-the-prairie-state.htm.

United Nations Department of Economic and Social Affairs (n.d.). "The 17 Goals,"
UNDESA, online at: https://sdgs.un.org/goals.

United Nations Statistics Division (n.d.). "SDG Indicators. Metadata repository" Available at:
https://unstats.un.org/sdgs/metadata/.

Valentine, P.C. (2019). "Sediment Classification and the Characterization, Identification, and
Mapping of Geologic Substrates for the Glaciated Gulf of Maine Seabed and Other Terrains,

Providing a Physical Framework for Ecological Research and Seabed Management." United States Geological Survey/USGS Scientific Investigations Report No. 5073. Available at: https://pubs.usgs.gov/sir/2019/5073/sir20195073.pdf.

White, J. (2021) a. "Statement to the Greater Rockford Airport Authority about Bell Bowl Prairie." September 23, 2021, online at: https://www.naturalland.org/wp-content/uploads/2021/09/Bell_Bowl_Prairie_statements_by_John_White.pdf.

White, J. (2021) b. "Bell Bowl Prairie and Chicago Rockford International Airport." Online at: https://drive.google.com/file/d/1mpzcHkbW5SCp10jQtw6TmVv63K1I9yhG/view.